PANAMA

WILDLIFE GUIDE

RAINFOREST PUBLICATIONS

Gato de agua, Nutria
Lontra longicaudis 60 cm

Greater Grison
Lobo gallinero, Tigrillo rosillo
Galictis vittata 50 cm

Gray Fox
Micho de cerro, Zorro gris
Urocyon cinereoargenteus 50 cm

Coyote
Lobo de pradera, Coyote
Canis latrans 80 cm

Capybara
Poncho, Capibara
Hydrochaeris hydrochaeris 100 cm

Bush Dog
Perro de monte
Speothos venaticus 70 cm

Central American Agouti
Ñeque
Dasyprocta punctata 50 cm

Paca
Conejo pintado
Cuniculus paca 70 cm

Nine-banded Armadillo
Armadillo
Dasypus novemcinctus 45 cm

Mexican Hairy Porcupine
Puercoespín norteño
Sphiggurus mexicanus 45 cm

Naked-tailed Armadillo
Armadillo cola de puerco
Cabassous centralis 35 cm

MAMMALS

Jaguar
Tigre, Jaguar
Panthera onca 150 cm

Puma, Mountain Lion
León montés, Puma
Puma concolor 110 cm

Jaguarundi (3 forms)
Tigrillo congo, Tigrillo negro (3 formas)
Puma yagouaroundi 65 cm

Ocelot
Manigordo
Leopardus pardalis 75 cm

White-nosed Coati
Gato solo, Zuto
Nasua narica 55 cmm

Crab-eating Raccoon
Gato manglatero
Procyon cancrivorus 60 cm

Kinkajou
Mico león, Cusumbí
Potos flavus 50 cm

Olingo
Olingo
Bassaricyon medius 43 cm

Cacomistle
Cacomistle
Bassariscus sumichrasti 45 cm

Striped Hog-nosed Skunk
Gato cañero
Conepatus semistriatus 40 cm

Long-tailed Weasel
Comadreja
Mustela frenata 20 cm

Hoffmann's Two-toed Sloth
Perezoso de dos dedos
Choloepus hoffmanni 60 cm

Brown-throated Three-toed Sloth
Perezoso de tres dedos
Bradyipus variegatus 60 cm

Northern Tamandua
Hormiguero bandera
Tamandua mexicana 60 cm

Giant Anteater
Oso caballo, Hormiguero gigante
Myrmecophaga tridactyla 130 cm

Baird's Tapir
Macho de monte, Tapir
Tapirus bairdii 200 cm

Wooly Opossum
Zorra roja, Zarigüella lanosa
Caluromys derbianus 25 cm

Common Opossum
Zorra común
Didelphis marsupialis 40 cm

Rufous Tree Rat
Gato bruja, Ratón marañero
Diplomys labilis 30 cm

White-tailed Deer
Venado colablanca
Odocoileus virginianus 130 cm

White-lipped Peccary
Puerco de monte
Tayassu pecari 100 cm

White-throated Capuchin
Mono carablanca
Cebus capucinus 40 cm

Geoffroy's Tamarin
Mono tití
Saguinus geoffroyi 28 cm

Central American Spider Monkey
Mono colorado
Ateles geoffroyi 50 cm

Western Night Monkey
Mono nocturno
Aotus zonalis 45 cm

Mantled Howler Monkey
Mono aullador
Alouatta palliata 50 cm

Montane Squirrel
Ardilla pigmea montañés
Syntheosciurus brochus 17 cm

Variegated Squirrel
Ardilla gris, Arilla negra, Ardita
Sciurus variegatoides 28 cm

Western Pygmy Squirrel
Ardilla pigmea de vientre rufo
Microsciurus mimulus 14 cm

White Tent Bat
Murciélago de tienda blanco
Ectophylla alba 4 cm

Davy's Naked-backed Bat
Murciélago espalda desnuda de Davy
Pteronotus davyi 5 cm

Greater Fishing Bat
Murciélago pescador
Noctilio leporinus 12 cm

Orange Nectar Bat
Murciélago nectívoro anaranjado
Lonchophylla robusta 7 cm

Nothern Ghost Bat
Murciélago fantasma norteño
Diclidurus albus 7 cm

REPTILES

Leatherback Sea Turtle
Tortuga canal
Dermochelys coriacea 200 cm

Green Sea Turtle
Tortuga verde
Chelonia mydas 150 cm

Hawksbill Sea Turtle
Tortuga carey
Eretmochelys imbricata 110 cm

Olive Ridley Sea Turtle
Tortuga Olivácea
Lepidochelys olivacea 80 cm

Spectacled Caiman
Babillo
Caiman crocodilus 150 cm

Loggerhead Sea Turtle
Tortuga carpintera
Caretta caretta 100 cm

American Crocodile
Cocodrilo aguja
Crocodylus acutus 280 cm

Green Iguana
Iguana verde
Iguana iguana 60 cm

Black Ctenosaur
Garrobo negro
Ctenosaura similis 60 cm

Jesus Christ Lizard or **Common Basilisk**
Moracho común
Basiliscus basiliscus 25 cm

Emerald Basilisk
Basiliscus verde
Basiliscus plumifrons 25 cm

Leaping Anole
Iguanita saltadora
Anolis lemurinus 8 cm

Turnip-tailed Gecko
Escorpión tobobo
Thecadactylus rapicauda 13 cm

Annulated Gecko
Gecko anulado
Gonatodes annularis 4 cm

Bromeliad Lizard
Lagartija de bromelias
Anadia ocellata 8 cm

♀

Rainbow Whiptail
Borrriguero arcoiris
Cnemidophorus lemniscatus 11 cm

Red-tailed Lizard
Lagartija coliroja
Gymnopthalmus speciosus 4.5 cm

♀

Four-striped Whiptail
Borriguero de cuatro rayas
Ameiva quadrilineata 9 cm

Red-bellied Galliwasp
Escorpión coral
Diploglossus monotropis 20 cm

♀

Panamanian Whiptail
Borriguero panameño
Ameiva leptophrys 13 cm

♀

Highland Alligator Lizard
Dragón de altura
Mesaspis monticola 9 cm

White-striped Climbing Skink
Limpia casas de raya blanca
Marisora unimarginata 9 cm

Mangrove Cat-eyed Snake
Culebra ojigata de manglar
Leptodeira rubricata 75 cm

Boa Constrictor
Bécquer
Boa constrictor 380 cm

Orange-bellied Swamp Snake
Culebra de manglar quilada
Tretanorhinus nigroluteus 90 cm

Black-tailed Cribo
Sabanera real
Drymarchon melanurus 300 cm

Oriole Snake
Mica
Spilotes pullatus 260 cm

Fire-bellied Snake
Culebra panza de fuego
Liophis epinephalus 80 cm

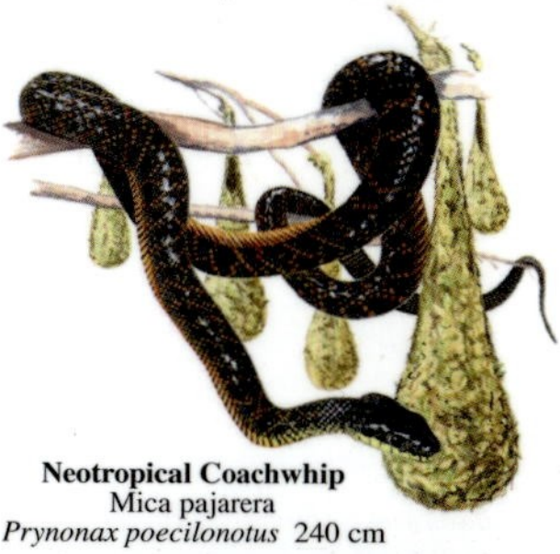

Neotropical Coachwhip
Mica pajarera
Prynonax poecilonotus 240 cm

Satiny Parrot Snake
Lora satinada
Leptophis depressirostris 180 cm

Red-eyed Treesnake
Culebra ojiroja
Siphlophis compressus 120 cm

Green Vine Snake
Bejuquilla verde
Oxybelis fulgidus 200 cm

Tropical Kingsnake
Coral falsa grande
Lampropeltis triangulum 180 cm

Coral-colored Centipede-eater
Cienpiesera de color coral
Tantilla supracincta 60 cm

Collared Glasstail
Cola de vidrio de collar
Urotheca decipiens 60 cm

Painted Keelback
Quillada pintada
Chironius carinatus 210 cm

Speckled Racer
Corredora pintada
Drymobius margaritiferus 130 cm

VENOMOUS

Eyelash Palm Pitviper
Oropel (forma amarrilla), Bocaracá
Bothriechis schlegelii 80 cm

Bicolored Coral Snake
Gargantilla
Micrurus mipartitus 110 cm

Choco Forest Pitviper
Vibora del Chocó
Bothrops punctatus 150 cm

Panamanian Coral Snake
Coral panameña
Micrurus stewarti 65 cm

Lansberg's Hog-nosed Pitviper
Patoco
Porthidium lansbergii 90 cm

Fer-de-Lance
Barbilla amarilla, Equis, Terciopelo
Bothrops asper 250 cm

Central American Bushmaster
Matabuey, Cascabel muda centroamericana
Lachesis stenophrys 390 cm

Brown Pelican
Pelícano pardo, OldJoe
Pelecanus occidentalis 119 cm

Anhinga
Aninga,
Anhinga anhinga 89 cm

Black-bellied Whistling-Duck
Pato silbador aliblaco
Dendrocygna autumnalis 53 cm

Magnificent Frigatebird
Fragata magnifica
Fregata magnificens 102 cm

Snowy Egret
Garceta nivea
Egretta thula 61 cm

Great Egret
Garceta grande
Ardea alba 99 cm

Little Blue Heron
Garza azul chica
Egretta caerulea 61 cm

Green Heron
Garza verde
Butorides virescens 46 cm

Striated Heron
Garza listada
Butorides striata 48 cm

Tricolored Heron
Garza tricolor
Egretta tricolor 66 cm

Wood Stork
Cigüeña americana
Mycteria americana 102 cm

Boat-billed Heron
Garza cucharón
Cochlearius cochlearius 51 cm

Purple Gallinule
Gallareta morada
Porphyrio martinicus 33 cm

Yellow-crowned Night-Heron
Garza nocturna cabeciamarilla
Nyctanassa violacea 61 cm

Gray-cowled Wood-Rail
Rascón montés cuelligrís
Aramides cajaneus 38 cm

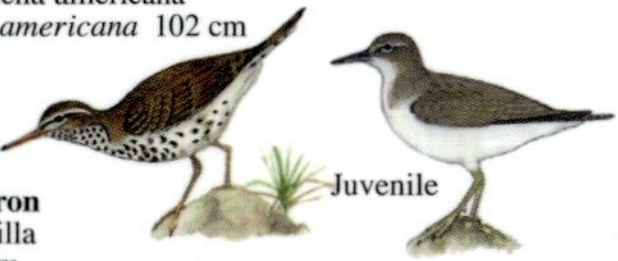

Spotted Sandpiper
Playero coleador
Actitis macularius 20 cm

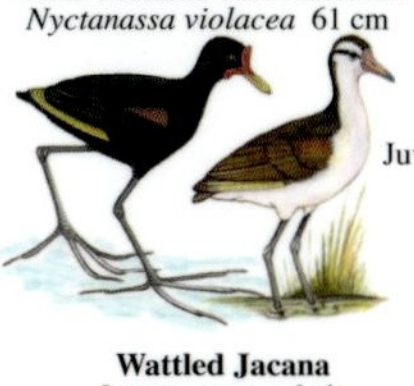

Wattled Jacana
Jacana carunculada
Jacana jacana 23 cm

Willet
Playero aliblanco
Tringa semipalmata 38 cm

Whimbrel
Zarapito trinador
Numenius phaeopus 43 cm

Black-necked Stilt
Cigüeñuela cuellinegra
Himantopus mexicanus 38 cm

Red-necked Phalarope
Falaropo picofino
Phalaropus lobatus 20 cm

Ruddy Turnstone
Vuelvepiedras rojizo
Arenaria interpres 23 cm

King Vulture
Gallinazo rey
Sarcoramphus papa 81 cm

Turkey Vulture
Gallinazo cabecirrojo
Cathartes aura 76 cm

Black Vulture
Gallinazo negro
Coragyps atratus 64 cm

Snail Kite
Elanio caracolero
Rostrhamus sociabilis 43 cm

Broad-winged Hawk
Gavilán aludo
Buteo platypterus 41 cm

White-tailed Kite
Elanio coliblanco
Elanus leucurus 41 cm

Black Hawk-Eagle
Aguilillo blanquinegro
Spizaetus tyrannus 66 cm

Harpy Eagle
Aguila arpia
Harpia harpyja 107 cm

White Hawk
Gavilán blanco
Pseudastur albicollis 58 cm

Common Black Hawk
Gavilán cangrejero
Buteogallus anthracinus 56 cm

Gray-lined Hawk
Gavilán gris rayado
Buteo nitidus 43 cm

Roadside Hawk
Gavilán caminero
Rupornis magnirostris 38 cm

Crested Caracara
Caracara, Cargahuesos
Caracara cheriway 58 cm

Osprey
Aguila pescadora
Pandion haliaetus 58 cm

Peregrine Falcon
Halcón peregrino
Falco peregrinus 46 cm

Collared Forest-Falcon
Halcón montés collarejo
Micrastur semitorquatus 61 cm

Yellow-headed Caracara
Caracara cabeciamarilla
Milvago chimachima 41 cm

Blue Ground-Dove
Tortolita azul
Claravis pretiosa 20 cm

White-tipped Dove
Paloma rabiblanca
Leptotila verreauxi 28 cm

Ruddy Quail-Dove
Paloma-Perdiz rojiza
Geotrygon montana 23 cm

Mealy Parrot
Loro harinoso
Amazona farinosa 38 cm

Red-lored Parrot
Loro frentirrojo
Amazona autumnalis 33 cm

Blue-headed Parrot
Loro cabeciazul
Pionus menstruus 25 cm

Orange-chinned Parakeet
Periquito barbinaranja
Brotogeris jugularis 17 cm

White-necked Jacobin
Jacobino nuquiblanco
Florisuga mellivora 13 cm

Yellow-crowned Parrot
Lora coroniamarillo
Amazona orchrocephala 30 cm

Brown-hooded Parrot
Loro cabecipardo
Pyrilia haematotis 23 cm

Violet-bellied Hummingbird
Colibrí ventrivioleta
Damophila julie 8 cm

Green Violet-ear
Orejivioláceo verde
Colibri thalassinus 10 cm

Purple-throated Mountain-gem
Colibrí montañés gorgimorado
Lampornis calolaemus 10 cm

Violet Sabrewing
Ala de sable violáceo
Campylopterus hemileucurus 15 cm

Blue-chested Hummingbird
Amazilia pechiazul
Amazilia amabilis 10 cm

White-tailed Trogon
Trogon coliblanco
Trogon viridis **26 cm**

Orange-bellied Trogon
Trogón vientrianaranjado
Trogon aurantiiventris 25 cm

Gartered Trogon
Trogón violáceo
Trogon caligatus 23 cm

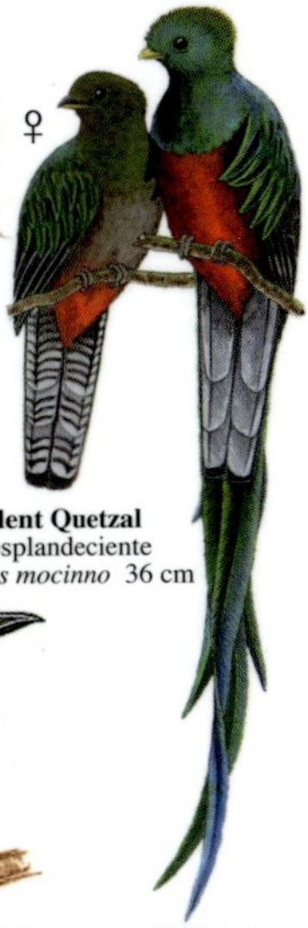

Resplendent Quetzal
Quetzal resplandeciente
Pharomachrus mocinno 36 cm

Broad-billed Motmot
Momoto piquiancho
Electron platyrhynchum 30 cm

Belted Kingfisher
Martín pescador norteño
Megaceryle alcyon 33 cm

Lesson's (Blue-crowned) **Motmot**
Pájaro bobo
Momotus lessonii 41 cm

Ringed Kingfisher
Martín pescador grande
Megaceryle torquatus 41 cm

Green Kingfisher
Martín pescador verde
Chloroceryle americana 18 cm

Yellow-throated Toucan
Tucán de Swainson
Ramphastos ambiguus 56 cm

Collared Aracari
Tucancillo collarejo
Pteroglossus torquatus 41 cm

Keel-billed Toucan
Tucán pico iris
Ramphastos sulfuratus 46 cm

Squirrel Cuckoo
Cuco ardilla
Piaya cayana 46 cm

Smooth-billed Ani
Garrapatero piquiliso
Crotophaga ani 36 cm

Common Potoo
Nictibio común
Nyctibius griseus 38 cm

Common Pauraque
Tapacaminos común
Nyctidromus albicollis 28 cm

Golden-olive Woodpecker
Carpintero verde dorado
Colaptes rubiginosus 20 cm

Crimson-crested Woodpecker
Carpintero crestirrojo
Campephilus melanoleucos 36 cm

Acorn Woodpecker
Carpintero careto
Melanerpes formicivorus 23 cm

Lineated Foliage-gleaner
Limpiafronda lineada
Syndactyla subalaris 18 cm

Barred Antshrike
Batará barreteado
Thamnophilus doliatus 15 cm

Fasciated Antshrike
Batará lineado
Cymbilaimus lineatus 18 cm

Red-faced Spintail
Colaespina carirroja
Cranioleuca erythrops 15 cm

Eastern Wood-Pewee
Pibí oriental
Contopus virens 15 cm

Scale-crested Pygmy-Tyrant
Trano-enano crestiescamado
Lophotriccus pileatus 8 cm

Spotted Antbird
Hormiguero collarejo
Hylophylax naevioides 10 cm

Social Flycatcher
Mosquero social
Myiozetetes simlis 15 cm

Panama Flycatcher
Copetón panameño
Myiarchus panamensis 20 cm

Great Kiskadee
Bienteveo grande
Pitangus sulphuratus 23 cm

Masked Tityra
Titira enmascarda
Tityra semifasciata 20 cm

Red-capped Manakin
Saltarín cabecirrojo
Ceratopipra mentalis 10 cm

Golden-collared Manakin
Saltarín cuellidorado
Manacus vitellinus 11 cm

Lance-tailed Manakin
Saltarín coludo
Chiroxiphia lanceolata 15 cm

Scrub Greenlet
Verdillo matorralero
Hylophilus flavipes 13 cm

Three-wattled Bellbird
Campanero tricarunculado
Procnias tricarunculatus 25 cm

Blue-and-white Swallow
Golondrina azul y blanca
Pygochelidon cyanoleuca 10 cm

Mangrove Swallow
Golondrina manglera
Tachycineta albilinea 13 cm

Gray-breasted Martin
Martín musliescarlata
Prongne chalybea 18 cm

Crimson-backed Tanager
Tangara dorsirroja
Ramphocelus dimidiatus 15 cm

Golden-hooded Tanager
Tangara capuchidorada
Tangara larvata 13 cm

Blue-gray Tanager
Tangara azuleja
Thraupis episcopus 15 cm

Scarlet-thighed Dacnis
Dacnis musliescarlata
Dacnis venusta 13 cm

Summer Tanager
Tangara veranera
Piranga rubra 18 cm

Yellow Warbler
Reinita amarilla
Setophaga petechia 13 cm

Prothonotary Warbler
Reinita protonotaria
Protonotaria citrea 13 cm

Red-legged Honeycreeper
Mielero patirrojo
Cyanerpes cyaneus 10 cm

Mourning Warbler
Reinita enlutada
Geothlypis philadelphia 13 cm

Chestnut-sided Warbler
Reinita flanquicastaña
Setophaga pensylvanica 13 cm

Northern Waterthrush
Reinita acuática norteña
Seiurus noveboracensis 15 cm

Black-bellied Wren
Soterrey ventrinegro
Thryothorus fasciatoventris 15 cm

Buff-breasted Wren
Soterrey pechianteado
Thryothorus leucotis 13 m

Rufous-breasted Wren
Soterrey pechirrufo
Pheugopedius rutilus 13 cm

Orange-billed Nightingale Thrush
Zorzal piquinaranja
Catharus aurantiirostris 15 cm

Long-billed Gnatwren
Soterillo picudo
Ramphocaenus melanurus 13 cm

Chestnut-capped Brush-Finch
Matorralero gorricastaño
Arremon brunneinucha 18 cm

Orchard Oriole
Bolsero castaño
Icterus spurius 18 cm

Slate-throated Redstart
Candelita gargantiplomiza
Myioborus miniatus 13 cm

Eastern Meadowlark
Pastorero oriental
Sturnella magna 20 cm

Baltimore Oriole
Bolsero de Baltimore
Icterus galbula 20 cm

Orange-billed Sparrow
Gorrión piquinaranja
Arremon aurantiirostris 15 cm

Crested Oropendola
Oropéndoal crestada
Psarocolius decumanus 38 cm

Yellow-rumped Cacique
Cacique lomiamarillo
Cacicus cela 28 cm

Great-tailed Grackle
Negro coligrande
Quiscalus mexicanus 43 cm

AMPHIBIANS

Red-eyed Leaf Frog
Rana arbórea calzonuda
Agalychnis callidryas 8 cm

Lemur Leaf Frog
Rana de hoja lémur
Agalychnis lemur 5 cm

Red-webbed Tree Frog
Rana arbórea patiroja
Hypsiboas rufitelus 5 cm

Pug-nosed Smilisca
Esmilisca chata
Smilisca sila 6 cm

Crowned Tree Frog
Rana arbórea coronada
Anotheca spinosa 8 cm

Rosenberg's Gladiator Tree Frog
Rana gladiadora de Rosenberg
Hypsiboas rosenbergi 8 cm

Horned Marsupial Frog
Rana marsupial de cuernos
Gastrotheca cornuta 8 cm

Hourglass Tree Frog
Rana arbórea variegada
Dendropsophus ebraccatus 4 cm

Masked Smilisca
Esmilisca enmascarada
Smilisca phaeota 8 cm

Emerald Glass Frog
Rana de vidrio esmeralda
Espadarana prosoblepon 3 cm

Granular Glass Frog
Rana de vidrio granulosa
Cochranella granulosa 3 cm

White-spotted Glass Frog
Rana de vidrio manchada
Sachatamia albomaculata 3 cm

Yellow-bellied Dart Frog
Ranita venenosa panza amarilla
Ranitomeya fulgurita 2.2 cm

Green-and-Black Dart Frog
Ranita venenosa verdinegra
Dendrobates auratus 4 cm

Red and Green Dart Frog
Ranita venenosa rojiverde
Oophaga granulifera 2.5 cm

White-lipped Foam Frog
Rana de espuma labiblanca
Andinobates fragilis 4 cm

Striped Rocket Frog
Ranita cohete rayada
Silverstoneia flotator 2 cm

False Dart Frog
Ranita venenosa falsa
Pristimantis gaigei 4 cm

Colombian Plump Frog
Rana oveja colombiana
Elachistocleis pearsei 7 cm

Oval Frog
Rana oválida
Elachistocleis ovalis 5 cm

Panamanian Golden Frog
Rana dorada panameña
Atelopus zeteki 4 cm

Illustrated by Mark Wainwright and Robert Dean

ISBN 188853842-2

Printed in the US on recycled paper.

www.rainforestpublications.com